20 THINGS YOU DIDN'T KNOW ABOUT BEETLES

DID YOU KNOW? BUGS

AF898645

DOUG BRADLEY

PowerKiDS press

Published in 2027 by The Rosen Publishing Group, Inc.
2544 Clinton Street, Buffalo, NY 14224

Portions of this work were originally authored by Arielle Chiger and published as *20 Fun Facts About Beetles*. All new material in this edition was authored by Doug Bradley.

Editor: Greg Roza
Book Design: Michael Flynn

Photo Credits: Cover Lukas Jonaitis/Shutterstock.com; series background freshcare/Shutterstock.com; p. 4 SKphotographer/Shutterstock.com; p. 5 ALFIZON/Shutterstock.com; p. 6 wonderisland/Shutterstock.com; p. 7 Charlotte Bleijenberg/Shutterstock.com; p. 8 Tomasz Klejdysz/Shutterstock.com; p. 9 Tricreative project/Shutterstock.com; p. 10 Tom_Deer/Shutterstock.com; p. 11 VVproduct/Shutterstock.com; p. 12 Prasetyo aj/Shutterstock.com; p. 13 Tomasz Klejdysz/Shutterstock.com; p. 14 Jason Patrick Ross/Shutterstock.com; p. 15 Independent birds/Shutterstock.com; p. 16 Colombe Photographie/Shutterstock.com; p. 17 Nyu Gallery/Shutterstock.com; p. 18 Mikhail Gnatkovskiy/Shutterstock.com; p. 19 Henk Bogaard/Shutterstock.com; p. 20 https://commons.wikimedia.org/wiki/File:Buprestidae_beetle,_FBNM.jpg; p. 21 Aleron Val/Shutterstock.com; p. 22 Daria Volyanskaya/Shutterstock.com; p. 23 George Trumpeter/Shutterstock.com; p. 24 Imogen Warren/Shutterstock.com; p. 25 (top) Mark Brandon/Shutterstock.com; p. 25 (middle) Michal Kimmel/Shutterstock.com; p. 25 (bottom) Nickolas warner/Shutterstock.com; p. 26 mehmetkrc/Shutterstock.com; p. 27 Conrad Barrington/Shutterstock.com; p. 29 Brum/Shutterstock.com.

Library of Congress Cataloging-in-Publication Data

Names: Bradley, Doug, 1971- author
Title: 20 things you didn't know about beetles / Doug Bradley.
Other titles: Twenty things you didn't know about beetles
Description: [Buffalo, New York] : PowerKids Press, [2027] | Series: Did you know? bugs | Includes bibliographical references and index.
Identifiers: LCCN 2025047120 (print) | LCCN 2025047121 (ebook) | ISBN 9781499457346 library binding | ISBN 9781499457339 paperback | ISBN 9781499457353 ebook
Subjects: LCSH: Beetles--Juvenile literature
Classification: LCC QL576.2 .B73 2027 (print) | LCC QL576.2 (ebook) | DDC 595.76--dc23/eng/20251208
LC record available at https://lccn.loc.gov/2025047120
LC ebook record available at https://lccn.loc.gov/2025047121

Printed in China

CONTENTS

THE MIGHTY BEETLES!

Would you believe there are more beetle species, or kinds, than there are species of any other group in the animal kingdom? They make up about 40 percent of the entire insect population. Beetles are definitely one of the strongest and most successful groups of animals on the planet. They've outlived many other animal species—including dinosaurs!

All beetles have six legs, two feelers (or antennas), and an **exoskeleton**. These amazing creatures can be very beautiful, helpful, and even entertaining. They also show off many wonderful colors, crazy horns, and odd habits.

Some beetles look like tiny works of art!

WIDE WORLD OF BEETLES

FACT 1

THERE ARE OVER 400,000 SPECIES OF BEETLES, WHICH IS MORE THAN ONE-QUARTER OF ALL ANIMAL SPECIES.

Beetles have the largest number of species of any animal in the world. Beetles vary greatly in size and shape. The rhinoceros beetle is the largest. It can grow to 7 inches (18 cm) long. The smallest beetles measure less than 0.04 inch (1 mm) long.

Male rhinoceros beetles use their long horns to dig and to fight other beetles.

IT'S HARD TO GET AWAY FROM BEETLES!

FACT 2

Beetles can be found almost everywhere, except in places where it's very cold. They live in fresh water, salt water, and on land. Some beetles even survive in hot springs. Many beetles live under the bark of trees both living and dead.

BEETLE LIFE CYCLES

FACT 3

SOME BEETLES LIVE LESS THAN A MONTH, WHILE OTHERS LIVE FOR MANY YEARS.

Some beetles live up to 12 years. Longhorned beetles eat the dry wood in which they live. Since wood isn't very nutritious, they spend most of their lives as **larvae**. Other beetles live no more than 3 or 4 weeks.

Longhorned beetles spend most of their lives as larvae. They live between 1 and 3 years.

LIFE CYCLE OF A BEETLE

ADULT BEETLES DON'T LIVE VERY LONG.

FACT 4

In time, beetle larvae change into **pupae**. This is called **metamorphosis**. Soon after the beetle becomes an adult, and its life's journey is just about over. Adult beetles rarely live more than 6 months.

BEETLE BODIES

FACT 5 BEETLES WEAR ARMOR.

A beetle's hard exoskeleton keeps it safe. It also has two sets of wings. The outer wings, called elytra, are hard. Elytra don't help the beetle fly. They help protect the beetle's body. They also protect the soft, inner wings, which are used for flying.

Beetles belong to the animal order Coleoptera (koh-lee-AHP-tuh-ruh). This name comes from the Greek words for "cover" and "wings."

Beetle larvae breathe through spiracles along the sides of their body.

BEETLES BREATHE THROUGH HOLES IN THEIR ARMOR.

FACT 6

Openings along a beetle's body allow the insect to breathe. These opening are called spiracles. Beetles take in air and make pumping movements to force the air throughout their body. The beetles use **muscles** to open and close spiracles so they don't lose too much water.

LET'S TALK

FACT 7

BEETLES TALK TO EACH OTHER WITH SMELLS.

Beetles can't see very well. They depend on **chemical** smells to **communicate** with each other. Female beetles produce and release chemicals called pheromones. Male beetles locate females by following the trail of the scent.

Male beetles use their antennae, or feelers, to follow a female's scent.

Even though beetles communicate by stridulation, people usually can't hear their calls.

BEETLES ALSO TALK USING SOUNDS AND VIBRATIONS.

FACT 8

Beetles make sounds by scraping their mouthparts together and rubbing their legs together. The process of rubbing body parts together to communicate is called stridulation. Beetles that live in a dead log create **vibrations** to communicate with beetles in other parts of the log.

BREAK IT DOWN

FACT 9

BEETLES WILL EAT JUST ABOUT ANYTHING.

Almost every product of the animal and plant kingdoms supplies some species of beetle with food. Beetles are decomposers. That means they break down and eat many things, including leaves, sticks, grass, carrion, animal waste, and much more.

Beetles work together on decomposing (taking apart and eating) a dead animal. "Carrion" is a dead, rotting animal.

FIGHTING BEETLES

BEETLES FIGHT EACH OTHER FOR FOOD, SOMETIMES TO THE DEATH.

FACT 10

Some beetles fight over food. These include stag beetles and dung beetles. Some even fight until one beetle kills the other! Ladybird beetles, or ladybugs, have been known to lie on their backs and play dead, just like they would if attacked by a **predator**.

Atlas beetles have three horns. Males use them to fight each other for food and mates.

FACT 11 WHEN THEY'RE IN TROUBLE, SOME BEETLES REALLY STINK!

When beetles are in danger, some stand on their head and release chemicals from a body part in their rear. The chemicals smell really bad and can turn other animals' skin brown. The horrible smell of the beetle's scent scares off predators.

Stink bugs, also called shield bugs, are beetles that make a stinky chemical when in danger.

Blister beetles, like this one, can spit poison at predators. The poison burns, allowing the beetle to escape.

SOME BEETLES CAN BITE, BURN, OR STING PREDATORS.

FACT 12

Some beetles have a more painful defense. Bombardier beetles spray a burning, stinging chemical from their **abdomens** with surprisingly good aim. Ladybird beetles, blister beetles, and soldier beetles store defensive chemicals in their blood and release them through their leg joints when attacked.

DUNG BEETLES

FACT 13

DUNG BEETLES MAKE GOOD USE OF POOP.

Dung beetles eat, lay their eggs in, and even live in the dung of other animals. That's another word for poop! Some dung beetles roll the dung into a ball and move it away for later. Some bury it in underground tunnels.

These dung beetles are feeding on a pile of horse dung.

Dung beetles form dung into balls so they can roll it to new places. They also lay eggs in these dung balls.

DUNG BEETLES HELP NEW TREES GROW.

FACT 14

When animals eat fruit, the seeds sometimes come out in the animal's waste. Dung beetles move the waste elsewhere and bury it. This puts the seeds safely in the ground. Dung beetles help the **environment** by planting new trees!

THROUGH THE YEARS

FACT 15

BEETLES AND DINOSAURS ONCE SHARED THE EARTH.

The earliest known beetles lived nearly 300 million years ago, along with dinosaurs. Once plant life started to grow, beetles began to reproduce at a much faster rate, surviving on the food that plants provided.

Beetle fossils have survived for millions of years. Fossils are the imprints or remains of plants and animals that lived long ago.

Some beetles, like the one shown here, can dive beneath the surface of water. They hunt for other creatures in the water.

BEETLES HAVE OUTLIVED MANY OTHER ANIMAL SPECIES.

FACT 16

Beetles have been around longer than many other kinds of animals. Their exoskeletons protect them from predators. Flying allows them to escape danger. They can dig holes and even swim, which allows them to survive changes in weather and climate.

FACT 17 SOME PEOPLE HAVE WORN BEETLES AS JEWELRY.

Beetle colors vary from dull brown to bright greens, blues, and purples. Some are even shiny. Fireflies are beetles that light up the night sky. In Europe and Asia, beetle wings have been used as ornaments for paintings and have even been worn as jewelry.

Beetles, also called scarabs, were commonly shown in the art and jewelry of ancient Egypt.

BEETLES ARE A HEALTHY FOOD SOURCE!

FACT 18

Did you know some people actually eat beetles? Beetles are high in fat and **protein**, which help people survive in places where there's not always enough food to eat. In some Asian countries, adult beetles and larvae are considered delicious treats!

NAME GAME

There are many beetles species, and many of them have some very colorful names! Many beetles have been named after objects, sounds, movements, places, and other animals. Christmas beetles were named in Australia for their glittery green color and because they suddenly first appeared around Christmastime. How do you think some of these beetles got their names?

CHRISTMAS BEETLE

STICK BEETLE

- ANT BEETLE
- TIGER BEETLE
- MONKEY BEETLE

- BARK BEETLE
- CLICK BEETLE

LADYBIRD BEETLE

GOLIATH BEETLE

- CARPET BEETLE
- JEWEL BEETLE
- DRUGSTORE BEETLE

FRIEND OR FOE?

FACT 19

LADYBIRD BEETLES CAN SAVE YOUR GARDEN.

Ladybird beetles, or ladybugs, are interesting to look at, but they're also very helpful. They eat other insects that are pests, such as aphids and mites. However, a few ladybird species are pests themselves because they eat crops.

There are about 150 species of ladybird beetles in the United States.

JAPANESE BEETLES CAN RUIN YOUR GARDEN.

FACT 20

Many beetles feed on the roots, wood, leaves, flowers, and fruit of living plants. Japanese beetles are very pretty, but the adults love to eat green leaves. The larvae love to eat roots. They can be a gardener's worst enemy!

BAD BEETLES!

Some beetles in the United States, such as the Asian longhorned beetle, are invasive species. That means that they somehow moved into a new area and hurt the local environment. The longhorned beetle eats and destroys many trees. It makes its nest inside piles of firewood. When the firewood is moved or disturbed, the beetles seek out and destroy new trees.

To protect trees and let beetles live in peace, never move firewood from one forest to another. This will stop the beetles from seeking homes in new trees.

Longhorned beetles don't actually have horns. Their antenna can reach about 4 inches (10.2 cm) long.

GLOSSARY

abdomen: The part of an insect's body that contains the stomach.

chemical: Matter that can be mixed with other matter to cause changes.

communicate: To share ideas and feelings through sounds and motions.

environment: The natural world in which a plant or animal lives.

exoskeleton: The hard outer covering of an animal's body.

larvae: Bugs in an early life stage that have a wormlike form. The singular form is "larva."

metamorphosis: The process of change that an insect goes through during its life.

muscle: One of the parts of the body that allow movement.

predator: An animal that hunts other animals for food.

protein: A nutrient in many types of food that the body uses to grow, repair tissues, and stay healthy.

pupae: Bugs that are changing from larvae to adults. The singular form is "pupa."

vibration: A rapid movement back and forth.

FOR MORE INFORMATION

BOOKS

Barclay, Maxwell V. L., and Patrice Bouchard. *Beetles of the World: A Natural History*. Princeton, NJ: Princeton University Press, 2023.

Jones, Richard. *An Anthology of Beetles: A Collection of Brilliant Beetles from Around the World*. London, UK: DK Children, 2026.

WEBSITES

Dung Beetle
animals.sandiegozoo.org/animals/dung-beetle
You can learn more about dung beetles at the San Diego Zoo website.

Types of Beetles
leafyplace.com/beetle-types-and-identification
Read about some of the coolest beetles in the world and see photographs of them.

INDEX